Algorithms for a New World

Alfio Quarteroni

Algorithms for a New World

When Big Data and Mathematical Models Meet

 Springer

Alfio Quarteroni
École Polytechnique Fédérale
de Lausanne (EPFL)
Lausanne, Switzerland

Politecnico di Milano
Dipartimento di Matematica
Milan, Italy

The basis of the English translation of this book from its Italian original manuscript was done with the help of artificial intelligence (machine translation by the service provider DeepL.com). The present version has been revised technically and linguistically by the author in collaboration with a professional translator.

ISBN 978-3-030-96165-7   ISBN 978-3-030-96166-4   (eBook)
https://doi.org/10.1007/978-3-030-96166-4

Cover illustration: Alfio Quarteroni

This Springer imprint is published by the registered company Springer Nature Switzerland AG.
The registered company address is: Gewerbestrasse 11, 6330 Cham, Switzerland

Acknowledgements

I wrote this book after a trip to Puglia (Italy) during which Rosella Santoro (the incomparable founder and soul of Il Libro Possibile festival) got me in touch with Claudia Coga of Dedalo. It was only because of Claudia's insistence and encouragement that I decided to write *Algorithms for a New World*—although the fall 2020 lockdown played a role. I would like to thank Claudia (and, indirectly, Rosella) for persuading me to take some time to muse on important topics that, otherwise, my academic commitments would have prevented me from exploring.

About the Book

How will the increasingly intrusive presence of artificial intelligence and computers capable of autonomous learning define our near future? Using a non-specialist language, this book will take us on a journey through mathematical models based on the knowledge of physical processes, and more recent models based on artificial neural networks. The purpose is to raise our attention to the challenges, opportunities, and dangers of a world regulated by mathematics and its algorithms and to show us how artificial intelligence and humanism can find a synthesis that benefits not only progress but also our well-being.

Contents

About the Author

Alfio Quarteroni is a mathematician and professor at Politecnico di Milano and the École Polytechnique Fédérale de Lausanne. Previously, he was a professor at the University of Minnesota in Minneapolis and a scientific director of CRS4. He is the founder and the director of MOX and the co-founder and the president of MOXOFF. He has received countless international prizes and awards, including the NASA Prize for his work in Aerodynamics, the Galileo Galilei International Prize for Science, the Feng Kang prize and the Euler Medal. He holds the Galilean Chair at Scuola Normale Superiore in Pisa and received an honorary degree in Naval Engineering from the University of Trieste. He is a member of the Accademia Nazionale dei Lincei, the European Academy of Science, the Academia Europaea, the Academy of Sciences of Lisbon, and the Istituto Lombardo Accademia di Scienze e Lettere. His work has been applied to medicine, earthquake geophysics, environment and climate change, aeronautics, and the oil industry. He led the mathematical study behind the design of Alinghi, the two-times America's Cup champion Swiss sailboat.

List of Figures

1

Epidemic

The sky is clear over the city, the air fresh and the light is bright on this day, the umpteenth, of a strange spring. Silent streets, closed shops and offices, green parks in bloom without the usual chatter of children. And all of us waiting for the usual afternoon bulletin, with its litany of cold numbers that might either give us some hope or unsettle us further. We suddenly realize that we are cursed by averages, curves and ratios that we would have gladly continued to confine to a hidden corner of our memory, the one linked to our—often difficult—relationship with mathematics.

It is an epidemic whose propagation front would be impossible to trace if we did not have, day after day, the numbers that represent and quantify it. It's the era of Big Data: so much data, heterogeneous in nature. The pandemic, the latest arrival, feeds them copiously. The data, by itself, says little. It must be interpreted, contextualized, made to live dynamically by projecting it into the process we are observing.

© The Author(s), under exclusive license to Springer Nature Switzerland AG 2022
A. Quarteroni, *Algorithms for a New World*,
https://doi.org/10.1007/978-3-030-96166-4_1

This data is collected in a non-homogeneous way. There is a time delay between one city and another, one region and another, because, for example, the processing time of the swabs is not the same, often different criteria, for example with different methods that could indicate positivity in one case or negativity in another. Or it is collected in an incomplete manner, due to a different approach to asymptomatic persons living with infected persons, or an insufficient testing capacity, or because of a precise political will (e.g. that of some countries that in the early stages of the epidemic decided not to communicate the data of patients in retirement homes—residences for the elderly). A similar consideration applies to the case of deaths: with what criteria should the data be attributed to the epidemic in the case of comorbidity?

It is difficult to extract significant information from raw data, like those of the daily bulletin, without a proper interpretation and contextualization, which is only possible through the tools of data analysis.

Even more crucial is whether this same data can help us understand what will happen tomorrow, i.e. whether past data can be used to predict future events. After all, that's what we're used to seeing mathematical models of weather prediction do: based on current conditions, they give us predictions about the days ahead.

A mathematical model, in general, is a "mathematical machine" that transforms an input into an output, in relation to a phenomenon that is being observed. To make this transformation possible, the phenomenon must be suitable to be written in mathematical terms, that is, through equations. This is an essential step that codifies the possibility that nature (as in the case of meteorology), or a biological and social process (in the case of an epidemic spread), can be represented with the language and tools of mathematics.

Let's take the case of the SARS-Cov-2 coronavirus discovered in China in late 2019. It spread over the world and generated the Covid-19 epidemic, which was declared a pandemic by the World Health Organization (WHO) on March 11, 2020. The equations go back a long time, roughly 250 years in Paris. Daniel Bernoulli, a member of a famous family of mathematicians and one of the fathers of the modern theory of probability, in order to advocate the cause of vaccination against smallpox, proposed a mathematical model by which he demonstrated that, if the entire French population had been vaccinated the general life expectancy would have extended by more than 3 years. A significant increase, given that at the time the average life expectancy in France was quite low.

Many years later, in 1927, the Scots William Kermack and Anderson McKendrick were trying to explain the rapid growth and subsequent decrease in the number of infected people in the London plague and cholera epidemics. They proposed a mathematical model destined to become for years a point of reference in the field of epidemic modelling. It's not far-fetched to say that the logical structure of the models that have been proposed to study the Covid-19 pandemic is just an extension of the one originally proposed by the two Scottish scientists. But how does one go from an epidemic (or even worse, pandemic) process to a representation in mathematical terms, that is, to a model? Let us proceed with order.

When it comes to an emerging virus (such as those of SARS type—Severe acute respiratory syndrome), everything becomes difficult because its characteristics are not known: virologists do not know its genetics, infectivologists do not know its virulence. Hypothetically speaking, since the entire human population has never come into contact with the pathogen and therefore has not had the chance to develop a specific immune defence, sooner or later it could become infected.

Thanks to this consideration, we could say that the entire population is susceptible, i.e. composed of individuals who, not having yet contracted the disease, can theoretically contract it. Once the pathogen has made the great leap from animals (probably bats) to humans and has infected its first healthy host (patient zero), the latter is potentially able to spread the disease in a virus-specific manner (e.g. SARS-type viruses are airborne, while HIV is passed on sexually). Susceptibles who come into contact with patient zero may in turn become infected, and in this way the epidemic spreads.

This is precisely the conceptual assumption of Kermack and McKendrick's original model: initially the population is divided into two categories, susceptibles (S) and infected (I). Of course, sooner or later some of the infected get rid of the infection, passing into a third category, the recovered (R). This term identifies those who no longer transmit the contagion (either because they are completely cured, or because the infection turned out to be lethal). Models of this type are called compartmental modes, the compartments referring to susceptibles, infected and recovered, and are consequently indicated by the acronym SIR.

Kermack and McKendrick first introduced a fundamental parameter, known as the basic reproduction number, or r_0, which expresses the average number of susceptibles that an infected person can infect in the first phase of infection. The larger r_0, the more virulent the outbreak. For Covid-19, estimates vary widely from place to place and time to time. In Italy, r_0 is estimated to have been around 2.5. It might seem a low value, especially if gauged against other viruses, but still lethal. The Spanish flu, by comparison, caused between 1918 and 1920 half a billion infections and tens of millions of deaths. It has been later calculated that its r_0 was around 2.1. In fact, if left to develop without any form of contrast, an epidemic with $r_0 = 2$ (just for the sake of an example, to make calculations easy) would lead to catastrophe: starting from patient zero, we would then have 2, 4, 8, 16 infected, that is 2^n after n passages. Let us remember that we are talking about average values: in reality there are infected people who fortunately do not infect anyone else, and others (the so-called super-spreaders) who infect a large number. Under these circumstances, without any prevention or containment measures, the growth curve of the infected would be exponential. As simplistic as this model may be, it does indeed match the real data. The incorrect mathematical understanding of the meaning of exponential growth has led the authorities in several countries to dramatically underestimate the effects of the epidemic in the early stages.

Fortunately, as the infection progresses, due to the introduction of containment measures, r_0 decreases from its initial value, and is generally referred to as r_t (transmissibility number, or coefficient). The critical value of r_t is 1: below this threshold each infected person infects less than one person on average, and the epidemic is destined to die out. With $r_t = 0.5$, for example, after n passages the average number of infected people is $\frac{1}{2^n}$, so on average it takes at least 20 contacts to generate a new infected person starting from as many as 1,000,000 susceptible people. If we want to contain an epidemic, therefore, we must ensure that r_t falls below this crucial value. If we have a vaccine, we can calculate that to be sure that the epidemic will extinguish itself we must vaccinate a fraction of the population greater than $1 - \frac{1}{r_0}$. For Covid-19, whose r_0 equals 2.5, the vaccine should be administered to at least 60% of the population. However, the variants that develop over time (the so-called

English, Brazilian and Indian variants, and who knows how many more) have the bad habit of raising the r_0 value. Consequently, the fraction of the population that must be vaccinated to reach the so-called herd immunity (or better, community immunity) in turn increases. For the English variant, with an estimated r_0 of 3.6, more than 72% of the population should be vaccinated! In the absence of a vaccine, we have to resort to containment measures, i.e. make sure that the potentially infectious do not stay in contact with susceptibles. A quick calculation tells us that if each individual came into contact with no more than 0.6 strangers on average, the Covid-19 infection (in its initial form) would stop. This is why we resort to isolation (quarantine), social distancing to the identification of new positives mainly through contact-tracing technology.

Once we have a good estimate of r_0 thanks to the data collection discussed above, how do we write the equations of an epidemiological model? In general, every model provides a simplified representation of reality. Sometimes it is very faithful, when it's just a matter of reproducing a non-living matter: how does the body of a car deform when it crashes against a wall? What kind of electric field is created in a lighted room? How does the temperature spread in a cooling room? These are just examples. But when biological and social processes are involved, the task is much more difficult and it is necessary to formulate certain simplifying hypotheses. In the case of a new epidemic, the first simplification is homogeneous mixing: each individual has the same odds of infecting anyone else, regardless of past contacts. This is of course a rather far-fetched hypothesis, as each of us has more frequent contact with family members, colleagues and friends, and we encounter more rarely the people who do not live near us or associate with different groups. But the effects of this assumption are not as serious as one might imagine. Moreover, in the simple model we assume that the incubation period (i.e. the time between infection and the onset of symptoms) and the course of the disease (from the onset of symptoms to recovery, or death) are constant. To make the model more realistic we should consider on the one hand the possibility of heterogeneous mixing, and on the other hand that the infected are removed from circulation at different rates, as a consequence of recovery, acquisition of immunity, withdrawal into isolation or death.

Let's stay with the simple case. Susceptible individuals S that leave the compartment of the infected I will increase the removed compartment R, according to the scheme S → I → R. The probability of transition from S to I is related to the average number of encounters that occur between the single susceptible and the individuals of the infected community. Analogous considerations are valid for the transition I → R. In the model we shall therefore conjecture an average time of permanence in phase S and an average time of permanence in phase I.

Let's indicate, for coherence, with the letters s, i and r the relative quantities of individuals in each compartment. That is, if n is the total number of the individuals of the population we want to examine (for example, n equals 60 million for the whole Italy), then $s = \dfrac{\text{size of } S}{n}$, $i = \dfrac{\text{size of } I}{n}$ and $r = \dfrac{\text{size of } R}{n}$. Under the assumptions we made, we can then deduce that the three quantities are linked by certain relationships that describe their variation in time. Solving these equations allows one to find, for each future time t, the number of susceptible, infected, recovered or deceased individuals at time t. Then the model could, for example, predict how many infected individuals there will be for $t = 30$, 60 or 90, i.e. in 1, 2, or 3 months' time.

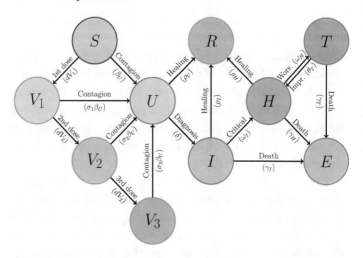

Fig. 1.1 The compartments of the SUIHTER model, an epidemiological model for the simulation of the COVID-19 spreading

In Fig. 1.1 we illustrate a modern version of a compartmental model, called SUIHTER, that has been developed to simulate the spreading of the SARS-COVID2 pandemic. The letters refer to the initial of the name of the corresponding compartment: S=Susceptibles, U=Undetected, I=Isolated at home, H=Hospitalized, T=Threatened, that is hosted in ICUs (intensive care units), E=Extincts, R=Recovered. V1, V2 and V3 correspond to the compartments of the vaccinated people with 1, 2, 3 doses, respectively (see "Modelling the COVID-19 epidemic and the vaccination campaign in Italy by the SUIHTER model", Nicola Parolini, Luca Dede', Giovanni Ardenghi, Alfio Quarteroni, *Infectious Disease Modelling*, Vol. 2, Issue 7, June 2022, pp. 45–63).

For the sake of illustration, we display in Fig. 1.2 the time variation of three different compartments in Italy during the first three epidemic waves.

In Fig. 1.3, we compare the number of hosted in ICUs in three Italian Regions (Lombardy, Piedmont and Sicily), normalized to 100,000 inhabitants.

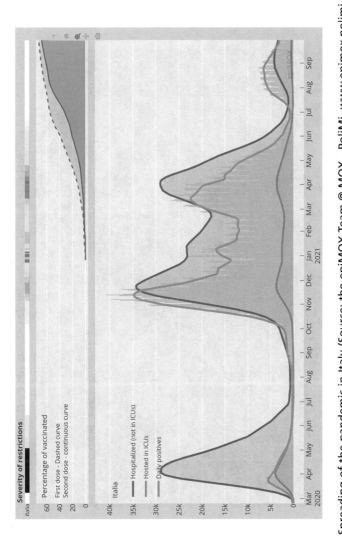

Fig. 1.2 Spreading of the pandemic in Italy (Source: the epiMOX Team @ MOX—PoliMi, www.epimox.polimi.it). The solid curves interpolate (by mobile averages) the daily data

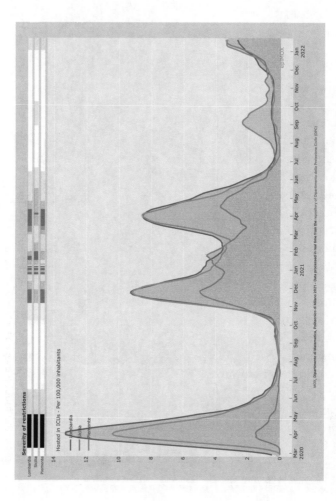

Fig. 1.3 Comparison among three Italian Regions along the first four epidemic waves (February 2020–January 2022) (Source: the epiMOX Team @ MOX—PoliMi, www.epimox.polimi.it). The solid curves interpolate (by mobile averages) the daily data

The crucial question is, of course, how reliable these predictions truly are. As is well known, many opinions have been voiced in recent months that challenge, sometimes in a virulent manner, the results of mathematical models published in major scientific journals and then spilled over to the media. I think a clarification is in order. Mathematically modelling a process such as a new epidemic is extraordinarily complex. We have already observed this before. In a model describing a deterministic physical process, once the data are known the model will produce a reliable solution, as accurate as one want. An epidemiological model, on the other hand, suffers from a great deal of uncertainty. Some of the parameters that constitute it are difficult to determine. For example, one parameter of the SIR model expresses the average number of infections per unit of time, another expresses the speed with which the transition from susceptibles to infected takes place. Yet another parameter indicates the average permanence time in compartment I. These are parameters that depend on the data: how many susceptibles, infected and cured individuals there are at the moment in which one wants to use the model to make predictions. Other parameters require some knowledge of the mobility between different cities: how many people travel from one city to another, in which conditions as regards distancing—on crowded trains, the underground or in private cars—or the differences of age brackets. As you can imagine, this information is difficult to obtain, also taking into account that it varies significantly over time, depending on the containment measures adopted. This inherent uncertainty requires a process of calibration of the parameters to the data, a very critical step especially in the presence of uncertain data. The intertwining of data and parameters, typical of epidemiological models, makes their predictive capacity much less effective than that of fully deterministic models. It'd be wise, then, to not expect accurate predictions from these models. Rather, we should expect plausible indications on different possible scenarios (from the worst case to the most optimistic one), within a range where the true quantitative answer will be found. I think that this should be considered the function of epidemiological models: to explore a collection of plausible scenarios (even if not quantitatively accurate) that provide the authorities with useful information for making decisions. (N. Parolini, G. Ardengi, L. Dede' and A. Quarteroni, *A Mathematical Dashboard for the Analysis of Italian COVID-19 Epidemic Data*, Int. J. Numer. Meth. Biomed. Engineering, 37(9) (2021)).

2

Retrospective

I decided to study mathematics almost by accident. When I enrolled in the bachelor course in mathematics at University of Pavia in Italy I didn't really know what to expect. I had completely insufficient background (coming from a commerce-oriented technical high school) and a vague idea that mathematics would play an increasingly important role in our lives. Above all, I had a great fascination for a difficult subject that was mysterious and feared by every student.

Falling in love was quick. Despite having a family background permeated with great pragmatism (I came from a rural context whose dominant values were those of hard labour), I immediately liked the abstraction of concepts and logical reasoning. The possibility of starting from hypotheses and deducing theses through rigorous and aesthetically beautiful demonstrations triggers a special thrill, like when you reach an unexpected goal or win a tough game.

While I became passionate about anything theoretical, towards the end of my degree I decided that I would

© The Author(s), under exclusive license to Springer Nature Switzerland AG 2022
A. Quarteroni, *Algorithms for a New World*,
https://doi.org/10.1007/978-3-030-96166-4_2

dedicate myself to the applicative aspects of mathematics. So much so that, fresh out university, I accepted an interesting job offer at an important national company. It didn't last long though, because after a few months I was offered to return to academia by applying for a job there. It was thus that I became a researcher, without ever having really thought of pursuing an academic career.

I became interested early on in how to use computers to solve mathematical problems of practical importance. For the layman, it should be noted that very often difficult mathematical problems cannot be solved "completely". In the sense that theory guarantees that there is a solution, but it does not help us to find it explicitly. In other words, we lack the "magic formula" that provides the solution.

This is what happens with the equations of epidemic spread that we met in Chap. 1. It also occurs (and even more so, since they are complex) with the equations underlying the mathematical models we use for weather forecasting, and in several other cases.

In all these situations, of course, mathematicians don't give up. They ask computers for help. Not because they know how to find the magic formulas that are unknown to mathematicians, but because they can actually solve "transformed" equations, that approximate the original ones but have the advantage they can be solved to the end.

This process is called, as a matter of fact, *approximation*. Even though we cannot describe it in detail (we would need to venture into mathematical technicalities) the idea is roughly as follows.

Let's start from our equation (though there can also be several), the one that describes a certain real process (the spread of the epidemic, the dynamics of the atmospheric variables, the pressure and speed of blood in our arteries, and thousands of others) and for which we are unable to provide a solution. We shall content ourselves with representing it in an approximate way, replacing it with other equations (many, sometimes very many) whose solutions, even if not exactly the original solution (the one we would like but we don't know how to find) are close enough to it. We say they *approximate* the initial solution.

On the surface, this operation doesn't seem to make much sense. We have replaced a single equation with many equations, and moreover, having introduced approximations, we have inevitably generated errors. However, transformed equations have two merits. The first, and crucial, is that now, thanks to computers, we can calculate their solutions in full. The other is that the approximate solutions obtained by the computer will be very similar to the initial exact one, as the errors generated will be small.

We do, however, need a mediator: between the approximate equations and the computer we will have to insert an *algorithm*, a computer program that will allow the computer, notoriously not (yet) endowed with own intelligence, to interpret our equations and solve them.

The fact that there are so many equations (sometimes tens or hundreds of thousands, sometimes even millions or hundreds of millions) would dishearten any human being, but not a computer, much less a supercomputer. Today on a laptop we can do a few billion algebraic operations (the four elementary operations of addition, subtraction, multiplication and division) in less than a second. Almost one billion billion, on the largest supercomputers in existence.

In this context, computers are therefore precious allies of mathematicians. They allow to find solutions to complex problems that are unobtainable otherwise, passing through the introduction of approximate solutions.

This process seems to disprove the popular belief that mathematics is *exact*. In fact, all (or almost all) solutions to mathematical problems found by computers are inevitably inexact, because they are approximate. We know that computers make mistakes in every single operation, since they are forced to operate with a finite number of digits. If you try to store the simple number 1/3, or 0.3333... on a computer, you will inevitably introduce an error, because it is physically impossible for the computer to find memory space for the infinitely many 3s after the decimal point. What matters, however, is that you have full control over these errors, that is, that you know in advance, even before solving a problem on the computer, how to *estimate* the error that will be generated. And in case this error is unacceptable, mathematics offers every tools to introduce approximate equations and make this error stay in the desired range. For example, if we are using a mathematical model to forecast tomorrow's weather, we can reasonably expect that the temperature in each neighbourhood of our city will be computed with an error not more than half a degree centigrade.

The conceptual process of describing a physical, biological, economic or social process with mathematical equations is called a *mathematical model*. As mentioned, the equations in the model must then be approximated in order to be solved by a computer. This process is called a *numerical*, or *computational, model*. The combination of the two, in sequence, identifies what is known as *scientific computing*. With scientific computing we aim at solving a real problem (we could call it *physical* for the sake of brevity, implying however that such a real problem can be originated by a natural, biological, industrial, financial, social, etc. process) by means of computers.

In Fig. 2.1 we display the classical roadmap of scientific computing: a physical problem is translated into a mathematical model, which consists of equations and of data generated by the physical model. The mathematical model is then approximated and generates a numerical model. The latter is turned into a computer program thanks to a

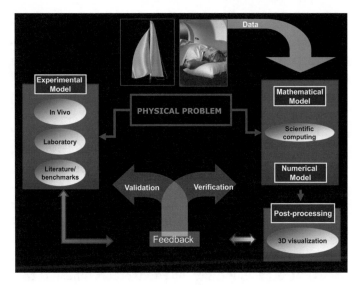

Fig. 2.1   The classical roadmap of scientific computing

suitable programming language, then solved by a computer. The results are then *verified* (how close are to the ideal solution of the mathematical model) and *validated* (how close are to the solution of the original physical problem). (A. Quarteroni, *The Equations of the Heart, of Rain, of Sails*, Springer Nature, 2022)

Scientific computing goes back in time a long way, and was born together with computers. Today we are used to solving problems of great complexity thanks to the existence of incredibly powerful supercomputers, as we saw above. But at the beginning of this adventure things were different. The most powerful computers of the time (1940–1950s) were not even comparable to what our mobile phone can do today. ENIAC, the first general purpose digital computer, delivered in 1945 to the University of Pennsylvania, could only perform a few thousand algebraic operations per second.

As you can imagine, progress in scientific computing goes hand in hand with progress in computers. More powerful computers encourage mathematicians to develop models for more complex problems. These generate numerical models with an increasing number of equations that require, in turn, more powerful computers. This is one of the many examples manifesting a perfect symbiosis between knowledge development and technological advancement.

In the summer of 1982 I visited the United States for the first time, thanks to David Gottlieb, an Israeli professor who, after inviting me to Tel Aviv University the previous year, suggested that I follow him to ICASE, at NASA's Langley Research Center in Virginia. ICASE was a world reference center for scientific computing, attended by the best researchers in the world. There, during the academic summer recess it was easy to meet the brightest minds from the most prestigious American universities. All efforts at that time were concentrated on solving problems that today we can propose to our students as projects in Master's courses. The equations of trans-sonic aerodynamics for two-dimensional airfoils, the study of one-dimensional conservation laws for non-viscous flows, the resolution of the Navier-Stokes equations for incompressible and non-viscous fluids in a fully periodic regime and with fully developed turbulence. The supercomputers of the time did not allow to do much else.

In the fall of 1990, when I accepted a full professorship at the University of Minnesota at Minneapolis, in the cold American Midwest, I was offered the position of Fellow of the Minnesota Supercomputer Institute. This was a state-of-the-art center that housed the most powerful prototypes of the Cray vector supercomputers, donated by Seymour Cray, the founder of Cray Research and graduate of the University of Minnesota. Their commercial value was of the order of tens of millions of dollars, and their performance at the time was not much higher than that of any common laptop we use today (cost: a few hundred euros). Yet, just 30 years ago, they exuded a sense of extraordinary power. They occupied a huge floor of the Minnesota Supercomputer

Institute, and the heat generated by their cooling system was conveyed down to a parking lot for employees and researchers. I remember that the thick blanket of icy snow that covered cars in the morning (in Minnesota it snows often, and winter temperatures are typically below -10 °C, with lows of -30 °C) had completely melted by lunchtime.

Already in those years problems of greater practical relevance could be tackled, also thanks to the Cray vector supercomputers. Just to give a few examples, it was possible to solve the full system of equations of fluid dynamics around real 3D configurations of commercial airplanes such as the Boeing 747, design the new Boeing 787 Dreamliner using scientific computing, develop realistic models for crash analysis, which allows virtual—as opposed to real—tests for the design of new cars.

It's interesting to note that Seymour Cray also had an eye for aesthetics. The vector computers of CRAY Research are splendid examples of design still today. When, a few years later in 1998, I accepted the Chair of Modelling and Scientific Computing at EPFL (École Polytechnique Fédérale de Lausanne, in Switzerland), I was surprised to find in the wide corridors adjacent to the lecture halls a permanent exhibition of attractively designed Cray supercomputers that had been used and then decommissioned. A legacy of a significant moment in the history of supercomputers.

Of course, today we can do more. For example, using the Marconi supercomputer installed at Cineca in Bologna, we can use mathematical models to simulate the effects of an earthquake on large urban areas. The aim is to simulate scenarios in seismic areas with vulnerable contexts to make predictive analyses of seismic damage to homes and infrastructures. Just for the sake of precision, the simulation for the Parthenon area in Athens requires solving over a billion equations, for several thousands time instants.

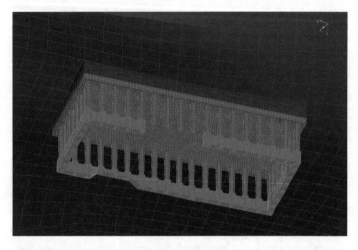

Fig. 2.2 The computational grid of the Parthenon in Athens (Courtesy: The SPEED Team @ MOX—PoliMi)

See Figs. 2.2 and 2.3 for an example.

These computations would be impossible without the power of the supercomputer Marconi, equipped with 347,000 cores and capable of almost 30 Petaflops at peak performance, or 30 million billion operations per second. Marconi ranked 21st among the 500 most powerful supercomputers in the world at the time of writing, but of course this is a fluid ranking that changes very quickly over time.

Another field of application of mathematical models is life sciences. One specific area in which models are achieving an extraordinary success is the simulation of the human cardiovascular system—what happens to our heart and arteries, where diseases with devastating effects on our lives arise. We must remind that deaths due to cardiovascular diseases account for over a third of all natural deaths in the Western world.

Mathematical models based on physical laws, those that regulate blood flow in the arteries or the heart's ventricles, together with the laws that underpin the deformation of

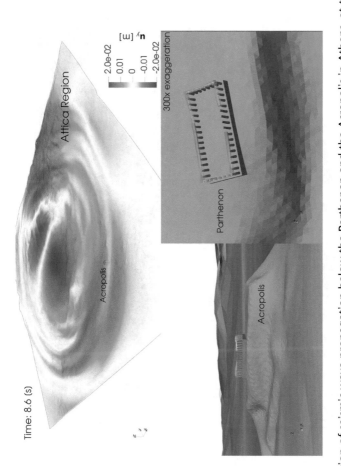

Fig. 2.3 Simulation of seismic wave propagation below the Parthenon and the Acropolis in Athens, at two different time frames: after 8.6 s (**a**) and 9.5 s (**b**) (Courtesy: The SPEED Team @ MOX—PoliMi)

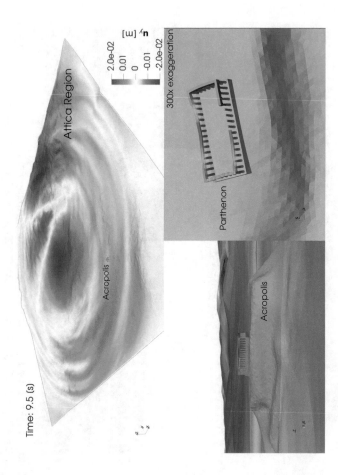

Fig. 2.3 (continued)

arterial walls or heart chambers, can represent a turning point in personalized cardiovascular medicine. For example, they can predict the outcome of therapeutic or surgical interventions on a specific patient. In these models, in fact, the parameters have a physical meaning, they are correlated by physical laws and can be adapted to a patient's changing conditions.

In Fig. 2.4 we display the computational grid used for the numerical simulation of a human heart.

In Fig. 2.5 we display the numerical simulation of the electric potential at three different instants of the heartbeat.

Figure 2.6 shows the evolution of calcium ions concentration in the early stages of the heartbeat. Calcium ions play the important role of "intracellular messengers": when calcium concentration inside the cell increases, the latter contracts, generating an active force. The beating of the myocardium is in fact driven by a calcium wave that propagates through the muscle tissue, acting as a signal (i.e. as a trigger) for muscle contraction. As shown in the figure, this wave first travels through the atria and then propagates into the ventricles.

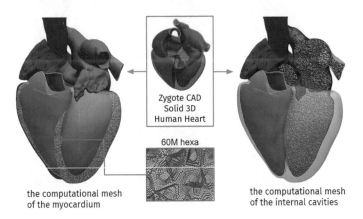

Zygote CAD
Solid 3D
Human Heart

60M hexa

the computational mesh
of the myocardium

the computational mesh
of the internal cavities

Fig. 2.4 A computational grid used for the human myocardium and for the heart cavities (Courtesy: the iHEART Team @ MOX—PoliMi)

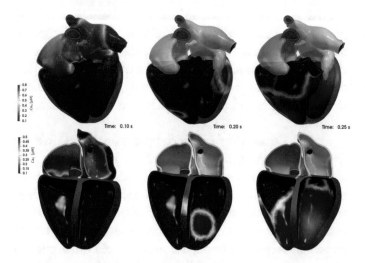

Fig. 2.5 The computation of the transmembrane electric potential at three different instants of the heartbeat (Courtesy: the iHEART Team @ MOX—PoliMi)

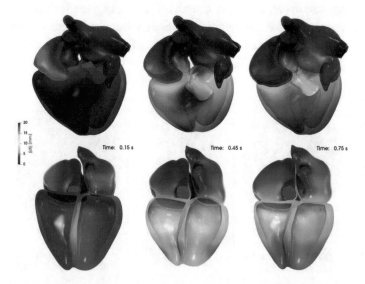

Fig. 2.6 The computation of the concentration of calcium ions in the early stages of the heartbeat (Courtesy: the iHEART Team @ MOX—PoliMi)

However, the final step towards any systematic use at the clinical level requires addressing a number of additional, critical issues. The first (and perhaps decisive) is the computational complexity inherent in the computer resolution of such models. The underlying equations are so many and so complex that they require a very long time to be solved. As an example, for simulating a single heartbeat, whose duration is about 1 s, it may take several days on a large cluster of parallel computers.

A further critical point regards the gaps in our scientific understanding of the underlying mechanisms. For example, the generation of molecular force in the myocardium due to the activation of transmembrane electric potential. Finally, it should be noted that some parameters that feed the mathematical model are difficult to determine for individual patients, as their quantification would require additional radiological exams that would not be justified diagnostically.

An alternative to that might be represented by data-driven models, which take advantage of artificial intelligence and machine learning algorithms. Their use alone, however, could be ineffective, as they are generally limited to the patient populations for which they have specifically been designed and trained. In fact, it is precisely from combining mathematical models based on physical laws and *data-driven* algorithms, such as those of machine learning and neural networks, that extraordinary, not even imaginable until a few years ago, opportunities arise. This topic is becoming increasingly relevant as time goes by, and deserves to be discussed in more detail. I will do that in the next chapters.

3

Interlude: The Revolution That Did Not Happen and the Revolution That Was Unforeseen

Fifty years ago I was fascinated, like all the boys of my generation, by the conquest of space. The explorations of the Russians and Americans with futuristic spaceships and fearless astronauts created a feeling of omnipotence. Humankind really seemed to have no limits. The conquest of the Moon in 1969 was experienced by everyone as a dream come true. Our satellite, which until then had mainly struck our imagination as the protagonist of romantic songs, suddenly became close and ... accessible. The *small step of one man* projected all of us, with our imagination, to great leaps into space, thus creating great expectations and paving the way in what seemed to all of us, at that time, the inevitable direction: the space revolution. Incredible adventures were being drawn in the minds of us kids, for we were convinced that in the following decades humankind would travel in space onboard comfortable spaceships with mesmerizing shapes.

A. Quarteroni, *Algorithms for a New World*, https://doi.org/10.1007/978-3-030-96166-4_3

Those hopes and illusions were, as it turned out, betrayed. Although exploration has continued with the Space Shuttle missions, and today, after a long pause, it has been revitalized thanks to Elon Musk and the never dormant projects of a future conquest of Mars, it is clear that exploration has never really come within reach of the common man.

The space revolution did not happen, at least in the form in which it was dreamed, imagined and promised.

The *real* revolution—the *computer* revolution, on the other hand, was not foreseen by anyone. At least not in the form and scale in which it happened. At the time I graduated, powerful computers were mainly the bulky *mainframes* in university computer centers and the big research labs of national interest in developed countries. Only in the 80s the personal computer revolution would allow everyone to have computing power available on their desks. The great equations that had marked the development of mathematics and physics in the previous century, such as the Laplace, Maxwell, Navier-Stokes, Schrödinger or Helmholtz equations—those that allowed to model physical phenomena mathematically—could finally be solved numerically, through approximation algorithms, thanks to low-cost CPUs and storage.

The computer revolution would go on to make an extraordinary impact on applied mathematics. With the availability of large computing resources, mathematicians on the one hand tackled increasingly complex problems, with the concrete hope that they could now solve them with more powerful computers. On the other hand, the door was opened onto creating new approximation strategies and new algorithms, according to paradigms more suitable to exploit vector and parallel computers. Classical algorithms were almost abandoned because of their low efficiency, for instance Jacobi's methods for solving large linear systems or Schwarz's method for solving the Laplace equation on regions partitioned into subregions, until they were rediscovered and generalized to the point that they are now the core of extraordinarily powerful algorithms for solving complex problems.

The second revolution that nobody had predicted was the one triggered by the introduction of the *World Wide Web*. Coincidentally, in 1993 I was head of the division of applied mathematics at CRS4, an international research center founded by the Nobel laureate Carlo Rubbia in Sardinia. At that same time the first Italian website, www. crs4.it, was being created and the first online European-scale newspaper, L'Unione Sarda (www.unionesarda.it), was being launched. The appearance of the first search engines marked a stepping stone in everyone's lives. I remember, towards the end of 1995, the excitement of using for the first time Altavista, a Digital Corporation browser that was later replaced by Yahoo, and the empowering feeling of being able to reach every corner of the world and every page of knowledge, which till then has been jealously guarded by the pages of encyclopedias. Those first searches, to tell the truth, required a certain patience. The answers were not always very fast, and the pages found were not always completely relevant. The process was, in fact, in its infancy, and would be followed by a phase of true explosion. The number of websites, and the corresponding number of pages, was increasing at a head-spinning pace: in 1996 there already were 180,000 websites, 2,700,000 in 1998, 30 million in 2001. Today there exist about 1.4 billion pages, with over 4 billion people with web access. More powerful search engines were needed to allow access to a network growing at this rate.

Larry Page and Sergey Brin, two doctoral students at Stanford University, had the idea of creating a novel algorithm, called *PageRank*, that allowed them to explore the web quickly and effectively. In September 1998 they founded Google and from that moment on our history would change. The birth of Google also marked the importance of developing new linear algebra algorithms to explore the web. In fact, the PageRank algorithm introduces an entirely original mathematical technique. Having been constantly developed over the successive years, it now enables to interrogate the web and receive immediate and relevant answers in a fraction of a second. New algorithms and new mathematics are needed to explore the web. This need would become even more stringent when social networks, first of all Facebook, appeared on the horizon, thus creating infinitely complex and intertwined structures, the graphs that connect people who share a link.

The web revolution and the social media revolution generate another revolution at the mathematical level, that of *Big Data*. Audio messages, photos and videos are exchanged at breakneck speed. Together with satellite data and the data collected more or less transparently from our mobile phones and the web pages we visit, they provide major international players an endless source to feed their gigantic databases. By 2025 it is estimated that we will be reaching the incredible amount of 100 Zettabytes (billion billion bytes)! It becomes necessary to recognize, manage, interrogate and process these mountains of data, and then generate models that translate into solutions, marketing proposals, profiling of individuals and groups, through the algorithms of artificial intelligence, more precisely machine learning and artificial neural networks. New mathematical developments are being imposed, and give rise to yet another revolution, the revolution of *mathematics of learning*.

I'm not using the word revolution by accident. We've seen how mathematical models deriving from the fundamental laws of physics provide powerful tools to observe the world, describe the most complex behaviours, guide the development and design in Industry 4.0, support doctors in understanding and treating disabling diseases such as cardiovascular diseases or those related to tumor growth, control and optimize production processes, even make predictive analyses.

Since we now have algorithms that allow machines to learn based on past experience, and as the latter is incessantly fed by an unlimited source of data (big data), we might be led to believe that machine learning can replace human learning. We might come to think that the great achievements of Newton, Einstein, Maxwell and Schrödinger may one day or another become obsolete and useless.

Is this the future we can expect? Will we end up relegating Einstein to the attic?

4

Artificial Intelligence, Learning Computers, Artificial Neural Networks

The question I'd like to answer, not necessarily in this chapter, is the following: *can we replace mathematical models with artificial intelligence*, more precisely with machine learning algorithms?

There's indeed a growing interest in artificial intelligence (AI). Almost every day we read articles in the media informing us of a sensational new application of artificial intelligence. We should stress that many of these articles are sensationalistic brags, and don't go into the specific merits of the discovery in question. Be that as it may, the expectation that's been created at a planetary level gives the discipline an almost magical halo, as has rarely occurred in the past for other subjects.

It'll take some time to answer the question. It is necessary to define first what AI is. For a mathematician, starting from the definitions is a primary need, almost like breathing. When we have a precise definition we have given ourselves clear reference points, and we have established a

© The Author(s), under exclusive license to Springer Nature
Switzerland AG 2022
A. Quarteroni, *Algorithms for a New World*,
https://doi.org/10.1007/978-3-030-96166-4_4

logical context in which to operate. With AI the first issues appear soon. In an abstract (and very general) way, we might say that AI is the capacity of computers to imitate cognitive functions of the human mind such as learning and problem solving. Therefore we could simplify by saying that a proof AI exists is the existence of "intelligent" computers.

The creation of the first intelligent computers probably dates back to the early 1950s, when, thanks to probabilistic methods, we built machines that took into account the probability of occurrence of a certain event. The term *artificial intelligence* was coined by John McCarthy, a Stanford professor, in 1956. It is not my intention to tell the history (albeit brief) of artificial intelligence. Instead I will limit myself to observe that today, when we think of AI, we are referring to two different paradigms to build intelligent computers. Traditional AI (we might refer to this one as classical), is made of classical programs, that typically use *object-oriented* languages or symbolic methods and will know how to make decisions (according to the logic of programming IF… THEN…: if a certain situation occurs, certain consequences follow). Among the numerous fields in which today classical AI is successfully applied we can cite artificial vision, recognition and use of written and spoken language, expert systems, robotics, self-driving cars and public transport.

Let us dwell for a moment on the application to driverless vehicles. Many car manufacturers have made prototypes capable of autonomous driving, even at medium speed on busy roads. These cars use a large number of sensors, cameras and localization systems such as the GPS. These are to all effects a learning system, more and more specific and refined, able to decide whether to brake, suddenly change course or adopt different driving modes depending on the context, after applying an algorithm that

Fig. 4.1 Autonomous driving cars—a noticeable instance of usage of supervised machine learning algorithms for artificial intelligence

takes into account the distance to an obstacle and the cruising speed. This represents a classical example of AI application (Fig. 4.1).

Another instance is, of course, robotics. Robotics is a discipline that develops methods enabling a robot (an inanimate mechanical medium) to perform specific tasks, by automatically reproducing what would otherwise be a human task. It's a field that was initially born within mechatronics, and in which today soft disciplines such as linguistics, psychology and biology join forces to create humanoid robots for alleviating human loneliness and substituting the desire for companionship.

A second paradigm is that of machine learning (ML), that is, a computer's ability to learn to perform a specific action without having been programmed with explicit instructions. It was also born in the 1950s: the famous name associated with it is that of the English mathematician Alan Turing, the man who helped decrypt the Germans' Enigma machine during the World War II. In 1950, in an article in Mind magazine, Turing formulated the famous question: *Can machines think?* To answer this question, Turing himself points out, first of all it is necessary to provide a precise definition of what we mean by *machine* and *thought*. Turing proposed a test, that took his name: we put a computer in a room, a human in another room, and eliminate any circumstantial element (voice, handwriting, etc.). An unaware subject is asked to guess, through some questions, whether they are talking to the person or the computer. The machine, of course, has the power to "cheat". The underlying idea is that if the ML algorithms are sufficiently powerful, the machine will become intelligent enough to pass as a human, and to not be identified by the third party (Fig. 4.2).

Let us step back to the present and consider autonomous driving. ML algorithms will allow our self-driving vehicle to see where pedestrians are, *recognize* those who are crossing the street, understand that they are human beings, and

Fig. 4.2 Statue of Alan Turing sitting in front of Enigma at Bletchley Park. On the right a picture of Turing (@Ian Petticrew-Wikimedia)

so on. They are called machine learning methods because there's no AI program, in the classical sense, that can tell whether the obstacle on the road is a human or an inanimate object. When a person has to decide whether the obstacle ahead is a pedestrian, or distinguish a cat from a dog, an ally from an enemy in a photo, the brain processes involved are complex and largely unknown. *Neural networks*, especially *deep neural networks*, are a mathematical tool used by ML in the attempt to replicate these (neural) processes in a simplified way. We shall deal with neural networks in the next chapter.

In spite of the extraordinary advancements made, it must be pointed out that autonomous driving will not become widespread any day now. Not for technological reasons, but rather for legal reasons, since there still are no laws regulating its use on public roads. ML activities have already achieved spectacular results in at least three main directions: image recognition, voice recognition (on many smartphones one can activate commands by speaking—think 'smart homes'), and text recognition and translation. We all have witnessed and benefitted from the recent incredible progress in these areas, every time we use Siri or Bixby, to name but two, or an automatic translator (to/from any language spoken on the planet), or finally when we tell an app to recognize an image.

But there is another area that is experiencing a very rapid growth, and is full of opportunities: the ability to profile potential customers (all of us) and put out tracing advertisement. This means that much of the current advertising targets the interests of the single internet user, whose needs and tastes are recognized through the analysis of their online searches. More generally, sophisticated ML algorithms are used to profile people's features, be they their preferences as consumers, as potential voters, or other. Of course, this use of ML has its own dangers and ethical and moral implications (I'll discuss these later).

New technologies are also revolutionizing the job market. According to some analysts, by 2030 AI will create 555–890 million new jobs worldwide, and by the same token 400–800 million people will change jobs. Needless to say, making predictions on such a large scale is extremely difficult. I think that when reading these figures we should not focus on the actual numbers, but rather consider their order of magnitude. We are, after all, talking about hundreds of millions of people whose professional destiny will be, in one way or another, marked by the hasty evolution of AI.

To the gamer reader, ML algorithms are regularly tested and released to create high-complexity matches in specific games, especially chess and backgammon. Competitive sports, including soccer, are no exception: top-level clubs now use the support of maths to optimize their performance. During every match of the Italian Serie A championship, the Spanish Liga, or the English Premier League, special video cameras record 20–30 times per second the coordinates of the ball and the 22 players on the pitch. There are about 10 million positional data generated over the course of a match. *Big data* algorithms are able to monitor all, and produce an incredible amount of useful indicators and evaluations. We are not talking about the statistics on ball possession shown on tv at the end of a match, but of tools capable of reconstructing every single elementary event in the game, like a shot, a pass, ball control, a dribble. All events whose effectiveness mathematics can analyze. Thanks to AI, it's possible to provide real-time operational indications regarding the aspects the coach and his staff have decided to monitor during the match itself. It's like having an extra virtual assistant. And as a matter of fact one such assistant already exists, called the *Football Virtual Coach*. This software platform was created by *Math & Sport*, a startup born from Politecnico di Milano, which the Italian Serie A has made available to all clubs since January 2020.

Medicine is another very promising area for ML. One recent and very timely example is a Chinese neural network trained with data from CT scans of patients' lungs, which seems to recognize the presence of the coronavirus in minutes, compared to the time it takes for the entire procedure of analyzing a swab. AI has already made its way into diagnostic imaging, in the automatic recognition of tumours and other diseases, or into smartwatches that provide heart rate or body temperature. This kind of data can be used to train very sophisticated neural networks that will likely alert doctors (remotely) and users to any critical situations. The recent SDSC-Connect 2020 conference, where researchers, innovators, industrial decision-makers and data scientists met to discuss how AI can foster discovery and creativity, has witnessed the emergence of new frontier areas, new territories that AI is successfully permeating. Further examples are the new drugs discovered thanks to *deep learning* applied to protein folding, the revolution in clothes design by big fashion companies due to generative approaches applied to garments, the creation of fragrances and aromas to complement the human senses of smell and taste, and also detecting *deep fakes* and putting the appropriate counter-measures in place.

As we might have guessed, for any AI application to work properly and reliably it is necessary for it to operate on sufficient and meaningful data. In fact, the main reason behind the rampant development of ML in the last two decades is the availability of Big Data and the possibility modern supercomputers offer to extract knowledge from it. The term Big Data generally indicates a collection of data so large in *volume*, generation *speed* and *variety* (or heterogeneity) as to require the development and use of special analytical techniques from maths and computer science to extract from it value and knowledge. Recently, the three above features have been joined by two more: *truthfulness*, in reference to their reliability, and *worth*, meaning the possibility of extracting value from such data.

In the last 30 years the volume of data in circulation has increased exponentially. In 1986 it amounted to 281 Petabytes (281 million billion bytes), in 1993 it was 471 Petabytes, in 2000 2.2 Exabytes (2.2 billion), in 2007 65 ExaBytes, in 2014 north of 650 ExaBytes and 250 ZettaBytes are estimated by the end of 2025. Simultaneously, we have witnessed the creation of new ways to produce, store and analyze this data, thus generating large and complex datasets in fields such as astronomy, meteorology, and biomedical research in general, particularly epidemiology, pharmacology and public health. At the same time, political and social trends such as the push for *Open Government* and *Open Science* have driven the sharing and interconnecting of heterogeneous research data across large digital infrastructures. These, in turn, serve as platforms for the development of AI, thanks to new algorithms for supervised learning, deep neural networks, stochastic minimization algorithms, and algorithms for visualizing complex data. I will mention these in the following chapter. These algorithms are programmed to "learn" from each interaction with new data. They have, in other words, the capacity to change themselves in response to new information introduced in the system.

Compared to a conceptualization of inquiry, that we might call Popperian, as the formulation and testing of a hypothesis, or the use of mathematical models derived from universal physical laws, the use of ML indicates a different (some might call it Baconian) understanding of the role of hypotheses in science. Theoretical laws are no longer seen as the drivers of the inquiry process. Rather, the empirical input (of data) is considered as primary in determining the direction research should take. The rise of Big Data therefore offers a new opportunity to recast the understanding of scientific knowledge as no longer only and necessarily centered around the theory.

5

A Bit of Maths (Behind Artificial Intelligence and Machine Learning)

Machine learning is the *discipline that gives computers the ability to learn without being explicitly programmed to do so.* This is one of the first definitions due to Arthur Samuel back in 1959.

In order to understand how the process of automatic learning works out, we can resort to the paradigm proposed by Tom M. Mitchell many years later, in 1997: *a program learns from an experience E with respect to a specific goal T (the task) and an assigned performance measure P if its performance with respect to T improves with experience E.*

As we can see, the process's formalization begins to take shape. To make this abstract statement concrete, let us take a classical example from image recognition. We want to write a program that, given an assigned image that we will call x, must return a binary number y: 1 if the starting image is a dog, 0 otherwise. We may suppose that an ML algorithm roughly works in the following way:

© The Author(s), under exclusive license to Springer Nature Switzerland AG 2022
A. Quarteroni, *Algorithms for a New World*,
https://doi.org/10.1007/978-3-030-96166-4_5

(E): we start from a *training set* that consists in a set of N images x_i (representing dogs or other animals) and a corresponding value y_i, (either 0 or 1), with i running from 1 to N;

(M): we introduce a possible *model*, that is, a function that transforms a generic image x into the binary value y. In symbols, $y = f(x,p)$, where p are values to be determined (typically called parameters) and f is the function that associates the pair (x,p) with the result y. The way f depends on x and p determines the chosen model;

(P): We *train* the model by choosing, among all possible parameters, those in correspondence of which the model, applied to the images of the training set, makes the smallest possible error. In other words, the values $t_i = f(x_i,p)$ differ the least from the true ones y_i (in the training set). One possibility to calculate this error is to add the squares of the discrepancies of t_i and y_i. Of course there exist other possible definitions of error, and the choice of such characterizes the performance measure P of our model.

A schematic example about the classical machine learning process is illustrated in Fig. 5.1.

For those who hate mathematical formalism, technicalities apart we may say that with ML we are proposing the model of a relationship between any image x and its semantic value (y, which indicates whether the image is a dog or not), where y depends on parameters chosen to optimize the performance of the model itself when applied to the training images.

At this point, having determined the numerical values of the parameters, our model is ready to be tested on any *new image*, one not already present in the training set.

Of course, changing context, we could now start from any set of inputs x (e.g. subjects affected by Covid-19) to generate an output y (e.g. the estimated number of recovery days).

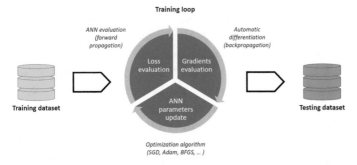

Fig. 5.1 The role of training and testing datasets in a machine learning process that uses an artificial neural network (ANN, see later). The parametric model underpinning the ANN is chosen upon minimizing a suitable loss function. The minimization (optimization) algorithm is based on a descent method that requires the evaluation of the derivatives of the loss function. Derivatives are calculated by automatic differentiation algorithms

There are several criteria for performing the experience phase (E). They are usually called *unsupervised*, *supervised* and *reinforced* learning. In unsupervised learning only the values x_i (not the corresponding values y_i) are included in the training set (the training is done "blindly"). In supervised learning both sets of values are included. In reinforced learning, the training set is not fixed a priori, but it changes and updates dynamically, according to the results and the context.

With a less mathematical, but more imaginative language, we may say that, in general, during the training phase of supervised learning we supply specific notions that allow to build a real database of information and experiences. In this way, when the computer faces a new problem, it doesn't have to do anything else than draw on the experiences already in its system, analyze them, and decide which answer to give on the basis of previously codified experiences. Unsupervised learning, instead, requires that the information inserted be not codified. This means the computer may access certain information without any example of how this information is used and, therefore, without knowing the results expected from the choice made. The computer itself will have to catalogue all the information in its possession, organize it and learn the results to which the information leads. Learning by reinforcement is the most complex of the three. The computer is supposedly equipped with systems and instruments able to improve its learning and, above all, understand the characteristics of the surrounding environment. This type of learning is typical, for example, of autonomous driving, where, as said, it benefits from an intricate system of cameras and supporting sensors.

It's not hard to understand that the choice of appropriate model f is crucial for the implementation of an effective ML algorithm. Clearly, many options are available in this regard. The simplest models provide a linear dependence of f on both x and p. Others provide a non-linear (polynomial) dependence on x, with polynomial coefficients represented by the parameters to be optimized. Then there are models of *decision-tree* or *random-forest type*, in which the choice of answer y is based on a sequence of yes/no choices, and the so-called SVM models (support vector machine). The latter fit in supervised learning. Starting from a training set whose elements are labelled by the category they belong to (dog/not dog, in the example), a model is built that assigns new examples to one of the two classes, thus giving a deterministic binary classifier. An SVM model represents examples as points in space, plotted as far as possible from one another if they belong to distinct categories. New examples are then plotted in the same space, and the prediction of the category to which they belong is based on the half-space in which they fall.

Several other models exist. The ones that have assumed a primary role are probably the models based on *artificial neural networks (ANN)*.

ANNs are mathematical instruments that try to emulate the behaviour of real neural networks—the biological ones—composed by basic nerve cells, axons and dendrites. The *mathematical neurons* of an ANN transform an input signal into an output signal. The elements that characterize each mathematical neuron are the parameters *p* of the model, as well as the transfer functions. These functions typically depend in a non-linear way on their arguments. Neurons are organized in arrays (or layers). Each neuron in a single layer receives an input from all neurons in the previous layer, and supplies an output to all neurons of the following layer. The first layer is the one giving the input values x_i, the final one will supply the output values y_j. In Fig. 5.2 we provide an example of artificial neural network.

In abstract terms, ANNs represent structures designed to simulate how the human brain analyzes and elaborates information. They are characterized by self-learning, which makes them capable to produce results that progressively

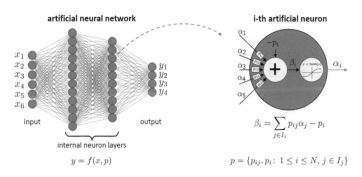

Fig. 5.2 An artificial neural network (ANN) together with its generic artificial neuron. This ANN is composed of two internal layers of artificial neurons. Every artificial neuron is acting as an input-output function that is the composition of a nonlinear sigmoid function with an affine transformation. The input-output map *f* transforms the input values *x* into the output values *y* and depends on a set of parameters *p* that are chosen upon minimizing a suitable loss function

improve as the available data's quantity and quality increase. The learning process happens during the training phase: given a dataset with known input and output values, a set of trainable parameters is calculated so to reconstruct precisely the underlying law, which will express the output as a function of the input.

Deep neural networks are characterized by a high number of internal layers. The different layers allow to represent complex and intertwined concepts. Suppose for instance we are interested in solving a classification problem, say, deciding whether a given image represents the 'person' category or the 'dog' category. Starting from an image characterized by a certain matrix of pixels, a deep neural network represents the image as a whole by decomposing it into simpler objects, such as edges, margins, corners or contours.

Another example: if we show a deep learning network a huge number of pictures of foods, the network will recognize whether a new picture represents a hot dog or not. If we show it images, videos and data collected from a car's sensors, it will know how to drive that car autonomously. And if we showed this deep learning network all the speeches of a former US President, it could come up with a completely plausible speech by that President (this has already been done, successfully).

Naturally, the architecture of these networks can vary a lot. In general, networks will depend on the specific type of problems that has to be solved. The ANNs used to categorize an image (dog vs. human) will be very different from those that have to, for example, calculate the risk of death of patients with ventricular fibrillation. ANNs are therefore interconnected layers of nodes (artificial neurons) able to "learn" to carry out classification or regression tasks, after they have been trained. There are several applicative fields in which ANNs have proved to be effective (often even more effective than human intervention). For instance, ANNs have already been applied to classify arrhythmias from single-lead ECGs (electrocardiograms) data of more than 50,000 patients, or to recognize the hazard of skin mole types. In either case they performed comparably to the experts in the sector.

Physics-based mathematical models and machine-learning (data-driven) algorithms represent two different paradigms to solve problems of real-life interest. For example, when using a physics-based mathematical model to simulate a "physical" problem (which could stem from physics, medicine, industry, sports performance, economy, or other fields), we use equations that underpin the "physics" at hand. Those are typically translating in mathematical terms some fundamental principles (like, e.g., the Newton's laws, the conservation principles of mass and energy, the thermodynamics principles, etc.). Those equations are invariant and apply to all problems of the same class. What characterizes a specific mathematical model is then the specific set of data, d, and, correspondingly, the solution u or some specific quantities that can be obtained from the solution, $y(u)$. This process is summarized in the upper part of Fig. 5.3. For instance, to simulate the distribution of the electric potential across the heart muscle of a given patient, the mathematical model is based (in particular) on the Maxwell equations of electromagnetism (that hold for every possible individual). The dataset describes the shape and the myocardium fiber structure of the heart of that specific patient, as well as the electric potential at a given instant (say, at the end of diastole). The output is the electric potential or the map of isolines of the electrical activation time on the epicardium (see Fig. 5.4 for an example). For the same problem, a data-driven algorithm (displayed in the lower half of Fig. 5.3) will use a family of data (for instance

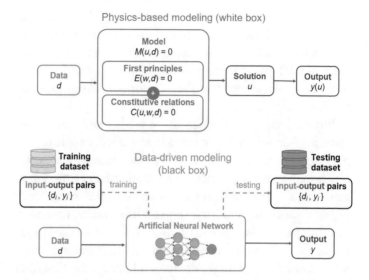

Fig. 5.3 Physics-based mathematical models (upper half) and data-driven (machine-learning) algorithms (lower half) are two different paradigms to simulate "physical" problems

corresponding to a cohort of patients) for the training and the testing of the artificial neural network. Then for a new patient, not comprised in the previous cohort, the ANN will be used to predict the outcome (in the case at hand the map of activation time isolines). No physical principles (henceforth no mathematical equations whatsoever) will be used in this second case.

One role still to be fully explored is how ML and ANNs can be used in *synergy* with the mathematical models based on physical laws, rather than as an *alternative* to them, thus strengthening a joint venture of extraordinary potential. In principle, there are many, diversified strategies to conjugate the science of Big Data with the science of universal laws. In many situations physics-based models are incomplete. This might depend on the lack of constitutive laws when modelling new materials, whose microscopic behaviour is not

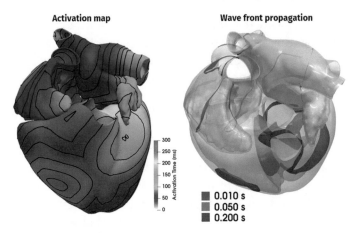

Activation map **Wave front propagation**

Activation Time (ms)
300
250
200
150
100
50
0

■ 0.010 s
■ 0.050 s
■ 0.200 s

Fig. 5.4 On the left, the map of the activation time of the transmembrane potential on the epicardium (the outer surface of the heart). Isolines connect points that are electrically activated at the same time instant. On the right, the activation front at three different time instants of a heartbeat (Courtesy: the iHEART Team @ MOX—PoliMi)

known. In these cases the use of ANNs, trained on these materials' behaviour when subjected to different types of stresses (the training data), can serve to find the missing laws at the phenomenological level. A similar situation occurs, for example, for epidemiological models, whose coefficients (average contagion time, average asymptomatic period) are not known theoretically. But they can be estimated thanks to ANNs that feed on observations, in case the epidemic has been present on the territory for several weeks (and lots of data are available to do the training). In other situations, the models based on physics come to the rescue of ANNs. This happens, for instance, when it's difficult to obtain the training input data. Think about models to be applied in the clinical field. If we wanted to use a neural network to predict the prognosis of a patient affected by a heart-rhythm disorder, we would need a large quantity of training data that may not be available (nor can we think of subjecting patients to additional tests with the only aim to feed the training set!). In cases of training-set deficit such as the one mentioned, we may use the physical model to generate, by varying its parameters, the data for the training set. It is an expensive operation, from the point of view of the computational resources, but possible.

Physical models are potentially very accurate and can produce solutions that adhere to the real solution. This is why they are called *high fidelity* (HF) models. Often, however, this accuracy comes at a very high cost. To get an idea of what this cost may involve, let us recall what we observed in Chap. 2. To simulate a single human heartbeat, a 1-s event in real life, the most accurate HF model we developed at Politecnico di Milano requires many hours, even days, on a supercomputer. In situations like this one, one usually develops reduced, or *low-fidelity* (LF) models, which allow to solve the problem in considerably less time than the HF model. The price to pay is the appearance of a slight approximation error. A reduced model offers an extraordinary benefit-to-cost ratio, due to the fact that the HF model still provides valuable information to the LF model. But how does one design a reduced model? Neural networks can do it well (accurately) and quickly (with reduced costs). It is precisely the synergy between physical models and neural networks that determines the HF-LF winning combination.

Finally, in phase (P) of the training, we can assume that the performance measure also takes into account how accurately the ANN-predicted output y satisfies the mathematical model (typically a differential equation, or a system of differential equations). The ANNs obtained in this way are "informed", or "aware" (they have been called *physics informed neural network*s, or PINNs). (en.wikipedia.org/wiki/Physics-informed_neural_networks)

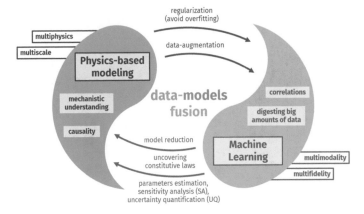

Fig. 5.5 The interplay between physics-based mathematical models and data-driven (machine learning) algorithms

These ways of using networks suggest that there is a lot of room for potential cooperation between the ML approach and the approach based on mathematical models derived from physical laws. This synergetic relationship is a harbinger of future developments that will, in my opinion, have a major impact from the viewpoint of mathematical applications.

See Fig. 5.5 for a schematic representation of the possible synergies between physics-based models and data-driven algorithms.

6

BIG DATA—BIG BROTHER (or, on the Ethical and Moral Aspects of Artificial Intelligence)

The main reason behind the impetuous development of ML in the last two decades is the availability of Big Data and the possibilities that efficient algorithms and modern supercomputers offer to extract knowledge from it. We mentioned in the previous chapter what extraordinary dimension this phenomenon has assumed and how Big Data motivated the creation of the highly efficient ML algorithms.

However, if we take a non-technical perspective, we must note that the dependence on big data collected by powerful institutions or private companies is starting to raise doubts and social concerns. The *digital divide* between those who cannot only access, but also use data is widening, and leads to a world of *data divide*. The privatization of data also has serious implications for the world of research and the knowledge it produces. The companies that hold the data usually only release what they consider of lower commercial value or requires public institutions to interpret it. The

© The Author(s), under exclusive license to Springer Nature
Switzerland AG 2022
A. Quarteroni, *Algorithms for a New World*,
https://doi.org/10.1007/978-3-030-96166-4_6

more expensive, or sensitive, or complex data, on the other hand, are jealously guarded.

The knowledge that can potentially be generated by any set of data lies in the extent to which this set can be linked to other datasets. Linking heterogeneous sets gives a high epistemic value to digital objects such as GPS positions or DNA sequencing data, to give two examples. Indeed, the aggregation of data from a wide variety of sources is often the premise for generating highly effective data analysis tools. Social media, government databases and research archives operate in a worldwide, interconnected and distributed network, whose functioning needs technological, computational and algorithmic skills. The disseminated nature of the decision-making process required for infrastructure development and Big Data analysis makes it impossible for any individual to maintain control over the quality, scientific significance and potential social impact of the knowledge produced.

Big Data analytics can therefore provide a platform that enables a distributed cognitive system. But what about liability? Many individuals, groups and institutions end up sharing the responsibility for the conceptual interpretation and social outcomes of specific usages of data. A key challenge for Big Data governance is to find mechanisms for allocating liability in this complex network, so that wrong and unjustified decisions—as well as outright fraudulent, unethical, abusive, discriminatory or misleading actions—can be detected, corrected and appropriately sanctioned. The great opportunities offered by information and communication technologies entail an enormous intellectual responsibility to understand and properly exploit them.

The implications of the use of Big Data and AI algorithms in sociology are no less relevant. Western self-centred societies are founded on the premise that no one can know our ideas or desires better than we do. This is why we elect ourselves, rather than those who govern us, to be in charge of our lives. The fear of many is that artificial intelligence will change this perspective, that ML algorithms may end up knowing us better than we know ourselves. The threat is that our employers or politicians may claim to know what their employees or citizens really want. If, for example, our governments gained access to these algorithms and knew what we think and like, they could meddle in our lives claiming it's for our own good. If this is the prospect, how can we expect to defend ourselves? By fortifying the shield protecting our privacy? By reducing the ability of others to exploit knowledge about us? A concrete example can help us understand the scenarios in which our privacy may be threatened. When we use an e-book reader, in all likelihood we are trading the convenience of reading a book in any circumstance with the possibility of privacy intrusion. The system can probably capture our underlined text, read our notes, estimate how often we read, assess our text comprehension speed (how long it takes to read a page).

Of course one could argue that algorithms able to know us perfectly can also be useful to us. For example by showing us which choices might make us happier or healthier, which professional opportunities are more suitable for us, which production or marketing strategies could make our company more profitable. The real problem is therefore to prevent these knowledge tools from remaining exclusively in the hands of others (and not ours). The alternative to erecting walls to protect our privacy is *algorithmic transparency*, i.e. the expectation that the structure, goals and decisions hidden in each algorithm be clear and explicit. We could require those who process our data to create knowledge about us should be legally bound to *return* that knowledge to us. In this regard, someone coined the slogan "nothing about us without us" in the age of artificial intelligence. That would be good practice, though difficult to enforce, especially for the privately owned web giants (the so-called Big Tech companies) such as Google and Facebook, for example. For some years now the Financial Times has been writing about the need to regulate technological platforms, for instance demanding that Big Tech be compelled to track the data used in their algorithms and be willing to clarify the algorithms' content to the public, but nothing significant seems to have been done in this direction.

There is also another threat that sometimes is seen. The threat that AI and ML algorithms, in the eternal attempt to perfect their learning capacity, end up invading spaces that we thought were exclusive territory of Homo Sapiens. *This is the fear that they will take over*. Homo Deus, an over-dramatized tale by Yuval Harari, tells what the world *might* be, rather than what it will be, when robotics, AI and genetic engineering are put at the service of the search for immortality and eternal happiness. The frightful risk is that the human race will end up making itself superfluous. Harari himself, however, acknowledges that no one can absorb all the latest scientific discoveries, no one can make predictions about the global economy in the next 10 years, and no one has a shred of a clue as to where we are heading in such a hurry. A narrative balance that relies on the physiological share of unpredictability that will affect humanity in the near future.

In fact, a possible limiting factor to the creation of computers capable of learning automatically is the human fear that machines could become too intelligent and take away their freedom of choice. According to a study conducted by researchers from Warwick Business School, the University of Plymouth, Radboud University in Neijmegen and the Bristol Robotics Lab, published in the journal *Frontiers in Robotics and AI*, it is reasonable to expect that an ML algorithm, hence a robot, will come to recognize a range of emotions and social interactions using movements, postures and facial expressions. The goal of the study is to create a robot that can react to the emotions of individuals living alone (as is often the case with the elderly), which might indicate stress or other particularly tough situations.

According to Arthur I. Miller, Professor of History and Philosophy of Science at University College London and author of "*The Artist in the Machine. The World of AI-Powered Creativity*", computers using artificial intelligence and ML algorithms can already create art, literature and music that could surpass human creations. They compose music that sounds more verse-like than Bach's, they turn photographs into paintings in the style of Van Gogh's Starry Night, they write entire books solely based on a subject, and they can even write screenplays. Miller explores the riches of computer-generated art, through artists and scientists who train AI to teach itself imagination, build robots that paint, write algorithms that create poetry. We may not understand the texts they create, Miller argues, but they will be able to explain them to us.

Mario Klingemann is an artist at the Google Cultural Institute in Paris. In his work "*Memories of Passersby*" he generates real-time portraits using neural networks running on a computer hidden inside an antique-looking piece of furniture, a combination between a mid-century modern piece of furniture and an old-fashioned radio. Klingemann argues that the work of art is not the images—for those disappear—but the algorithm that creates them. And he says human creativity is limited, whereas the creativity of tomorrow's computers will not be.

The technological revolution induced by AI is likely to be the fastest we human have ever witnessed. According to a pre-Covid-19 estimate by PwC (with the pandemic, all estimates risk becoming a rhetorical exercise), global GDP will increase by 16 trillion dollars by 2030. This will also bring huge challenges in terms of jobs. The single assembly line job is being taken by robots: we all expected this (and it is already a reality since Industry 3.0). But the impact will not only be in factories: truck drivers, car drivers, telemarketers or customer care workers, even radiologists and other specialists could be gradually replaced by artificial intelligence in the next 20 years. The most repetitive and routine jobs will probably be wiped out. And maybe in 20–30 years' time our children and grandchildren will be grateful to us for it.

What is more serious than the loss of work is the loss of meaning. Work's ethic leads us to think that work is the reason we exist, that work and success in your job defines the meaning of your life. Artificial intelligence will provide new analytical tools to stimulate and enhance the creativity of scientists, artists, musicians and writers. It will work alongside humans to boost their creativity by making them more original, perhaps more unique.

There is however another Cartesian axis, in addition to the creativity axis, along which AI will be able to help but not replace us humans. An axis along which we can align feelings of compassion, empathy, joy and sorrow. We can be sure that when AlphaGo, a software for the game of Go based on deep neural networks and developed by Google DeepMind, defeated the South Korean champion Lee Sedol, it didn't feel any emotion, while poor Lee was dejected. Here, emotion and joy are feelings that AI and ML algorithms can never feel (well, maybe…).

We can differentiate ourselves with those jobs that are both compassionate and creative, using and leveraging irreplaceable brains and hearts. Social workers and teachers who help us and our children find our way to survive and thrive in the AI age, nurses and doctors who not only care for us but also make us feel less alone through the tremendous Covid ups and downs, and who knows how many more. Bronnie Ware, an Australian author, in one of her books talks about the wishes and regrets of people on their deathbeds. When facing death, she argues, no one regretted not working hard enough, only that they did not spend enough time with their loved ones or make them realize how much they loved them. States of mind that help us understand that we have a territory to protect jealously from the intrusive front of AI. A territory to protect, within which to seek a model of coexistence of human beings and artificial intelligence, in the awareness of our absolute uniqueness.

Printed in the United States
by Baker & Taylor Publisher Services